SUBSIDENCE

ALSO BY JULIA JOHNSON

Naming the Afternoon (2002)

The Falling Horse (2012)

SUBSIDENCE

Julia Johnson

groundhog
POETRY PRESS

2016

Copyright © Julia Johnson 2016

All rights reserved

Library of Congress Control Number: 2016944946

ISBN: 978-0-9976766-2-4

Printed in the United States

Published by

Groundhog Poetry Press LLC
6915 Ardmore Drive
Roanoke, Virginia 24019-4403
www.groundhogpoetrypress.com

The groundhog logo is the registered trademark ™ of Groundhog Poetry Press LLC

For K. M. Y.

CONTENTS

I

Pointe-aux-Chenes, 1896	3
Prediction	4
River Diversion	5
Grand Isle Lower School Library	6
Trinity Island	7
Evolution	8
The Fault Plane	9
In the Holocene	10
Lonely Life	11
The Time We Were Lost	12
Subsidence I	13

II

Displacement Surface	17
Patterns	18
Representation of Figures	19
Elbow of Capture	20
Bird's Eye View	21
Sequential Land Leveling	22
The Effectivity of Processes	23

Bedtime	24
Error of Closure	25
Pulsation Tectonics	26
Morphology	27
Rectangular Coordinates	28
Circle of Ascension	29

III

Mother	33
Loneliness of a Kind	34
Golden Meadow	35
Nighttime in Sequence	36
Gulf of Lions	37
If We Are Very Still	38
Raccoon Point	39
At the Coastline	40
Holding the Pose	41
A Worry	42
The Flashlight Shines Bright Into the Living Room Window	43
Cheniere Caminada	44
Island Road	45
History of Wetland Loss	46

Tide Gauge	47
Unmasking of Fault Displacement	48
To Step Lightly	49
Phosphorus	50

IV

Grand Isle	53
Mangrove	54
Enigma	55
Barrier Island, Offshore Bar Theory	56
Barrier Island, Submergence Theory	57
Barrier Island, Spit Accretion Theory	59
Isle de Jean Charles, 1976	60

I

Pointe-aux-Chenes, 1896

In the murky water we hold a black snake.

He slips like a sash through our fingers.

We plunge him back in to see

if he can breathe in dimness.

We trade him back and forth, walk backwards

with him on the planks, and take his pulse.

When he dries in the sun we put him back

in an old cloth pocket. He looks like a moving

necklace now, just underwater. The sun is hot.

We squint until we can only see edges.

We think of hunkering if it rains.

Our snake is little in the pocket.

We want to lose him the next time there is dark.

Prediction

We can predict eclipses of the sun and the moon.

We can predict weight gain.

We can predict floods and droughts.

We can predict the spread of dementia.

We can predict the outcome of pregnancies.

We can predict adult height.

We can predict genetic disorders.

We can predict the wind.

We cannot predict hurricanes.

We cannot predict the extinction of religion.

We cannot predict analgesia provision.

We can predict how the temperature of the ocean currents

on one side of the globe

affect winter temperature on the other.

We cannot predict the moment of revolution.

River Diversion

We live on the plain.

On the plain we have been

hydrologically isolated from the river.

From the river, by containment,

levees for nearly a century.

For nearly a century, ensuing lack

of fluvial sediment inputs.

Sediment inputs, natural submergence,

process high coastal land,

loss rates, land loss.

Rates controlled river diversions.

River diversions have since been constructed

to reconnect the marshes of the deltaic.

The marshes of the deltaic, plain with the river,

the river pulsed, diversion sediment delivery, sea-level rise.

Grand Isle Lower School Library

We are instructed to look at a geological map of France, identify the mountain ranges. The teacher has a hand in a cast and she is chewing her tongue. She's not our teacher but she's there to tell us about the map. We are standing in a semi-circle around her. She doesn't let us get too close to the map. She talks about the geologist who prepared the map, that he also came up with the Offshore Bar theory of barrier islands in the 1850s. She wipes one corner of her mouth with the back of her hand.

Trinity Island

You are driving your small boat into the crease. The school is underwater now. You ride by the store with the cashier whose face is always in profile. You are trying to remember the teacher's story yesterday of her much younger ex-husband, an accountant with dark hair, who lives on the mainland, with a kind of heart condition. You are trying to remember the heart condition she said he has. Your uncles are situated in the highest house. One is playing cards, the other is sawing through the roof. The two dogs in your lap are licking their paws.

Evolution

In our splendid grass hut, we hid kindness.

Beyond the blue wall, we captured the white lies.

On bark, we scraped friendships.

Our wives were lost in hilltops.

We kept the actors on the stage.

We stole trees from the fields.

Our heads were held in hands.

The Fault Plane

You woke and found the switches pulled.

You fell onto the ratchet. Your head was a knob.

You were remarking and the side of your face

was a painting and it was a cement façade, a cart, a white sky.

You asked me if I was heavy, gave me simple cues as if recorded,

blocked signs, a pulley, a little mask to fit over my finger.

The flowers were set to full bloom.

Even the crease of your leg felt the hail.

You rode in your own steel cart down the driveway

and you were gone.

In the Holocene

Today, I read in one book of twenty-seven epochs—
but I count more than this on another list—subdivisions of the geologic timescale—
each longer than an age, shorter than a period.

Today, I remember we are living in the Holocene,
the quaternary period, *entirely recent*.

Today, we are in recent time,
in the interglacials, the deep sea cores tell us the oxygen isotope ratio.

Today, I look at a graph of climate change going back one million years ago.
Before the Holocene there was global warming.

Today, we know in the late Holocene, spears were replaced by bow and arrow.

Today, a Dutch scientist in New Orleans looks at the sedimentary record of the Louisiana coast, the Holocene sea-level change, connections with paleoclimate records. He will try to better understand source-to-sink sediment flux,
shelf-edge formation.

Lonely Life

The unicycle's lonely life

she woke up thinking

her grass skirt skirted skin

the green girls got upset

but this was nothing new

she ignored them always

the heart-shaped hand was hers

as she held it on the bus

her mother punished her

sent her to the boys playing

at the canal catching minnows

the needles on the sewing

table were with other needles

she thought

she found a book and read

the book slowly then once quickly

her face in the mirror was boring

The Time We Were Lost

We set out for the amazing forest

with the huge trees, cramped into a red truck.

Holding on for life was our task, our faces

pulled back by the wind.

We shared a backpack, a camera.

Our map was smeared.

When we stopped, we put our children

together in a high field,

fed them small pears, a teacher watched them.

The road was underwater, so we walked.

We waded closer to the sound of a train,

a thunder, one-Mississippi, two-Mississippi and then, another.

Subsidence I

A shock of flood. My father waits in the living room.

His house is gently shaking in the dark room pan. The pilings

are pins. My father waits every day for the newspaper, the mailbox not truly

floating, its red flag a sail. My father waits in the living room.

The radio with last week's battery. The cat jumps from the wide camper's room

to the tree, then deck.

The motors are killed.

The motors are killed.

My father waits in the living room.

II

Displacement Surface

Her mouth is made of sod and turned up.

Sad mouth.

She likens herself to a mouse.

She sits like a curve.

Her arm like a bud.

She is habit to her own self.

Her wind is different from ours.

She rides a low bicycle in the day.

Her mother is dressed like a bride.

Patterns

The seamstress describes the bodice and the rows

of seamstresses in training

fold their fabric in ink to mark the pattern.

Their own green satin billows on the table next to each of them.

She goes on to describe the shirt.

We are seamstresses describing the bodice.

We fold our fabric in half.

We work the fabric with chalk.

We will wear our clothes once completed.

Our shape is average and we have pinned the excess.

Representation of Figures

Look for the message. Early dial. The park is a patch on the map.

The birth snaps under the breakfast barn. Here we are.

We weld true to true. We hope for scopes.

The pulse is a range of figures in wind.

Elbow of Capture

The apple held high, a legitimate and carved frown on the plate.

We want early morning duds, the handkerchief in pocket, a tree-shaped fold.

We are waiting our turn in a pile, like cards of a kind.

We hand over this single, shifting and centered house in the head.

Little house in the head. The bravest go in.

Bird's Eye View

Passive appearance, lowlands:
intensity masking the region's
geological processes dynamic.
Mississippi River Deltaic Plain
and Chenier Plain lie above
a sediment-filled trough,
the salt basin. Was it 225 million
years ago or 226, Pangaea pulled apart?
Thickness of rock.

Tectonic movements, rivers flow:
particle by particle sands,
silts, clays, carried and dropped,
the weight of sediment pushed
down the crust.
The trough and the gulf deepen.

Sequential Land Leveling

I catch a stale rig in my hand. It is crooked and left rust.

I told my cousin to wait for the same thing to happen to him.

Burning sleeves in my dreams are always turning into tunnels

but my college friend knows I make up stories.

I head down the street where they are mopping up after the parade.

We slip into our white suits with shields for eyes.

I wrap up the Maidenhair Tree in brown paper after cutting it down.

The leaves are replicated on my living room walls.

On a visit, I notice my doctor's head has been painted with thick curls.

The Effectivity of Processes

We have found extraordinary sheep in the green hills,

white on distant shapes,

levels of paint in the seamless pond.

We have found perfect fuses to limit the light.

We have discovered cross-hatched saddles in the road.

We have come to this, blazoned.

We have found safety in what is kept distant.

Our hands are tied to the boat.

We have found a mask the size of someone else's face,

a map outlined in red, the shape a sandal.

We have found the first direction now: an arrow: *nine miles ahead*.

Bedtime

In our beds we keep

the largest oyster shells.

We cover them with sheets

while we're away. Our mother

is too small to wear our uncle's coat.

She pulls a blanket around her shoulders.

We go out into the night with our white

boots and lunch bags. Our mother lies

about our age to the man at the dock.

We slip under the rope, hop

onto the green dragger. We lift

the wings, horizontal spread,

hose the nets, then lower them down

for midwater trawling.

Error of Closure

I will live in a container, small and red and square.

I will wipe my hands on the white dishtowel.

I will lie on the cot.

I will turn old.

The cars will round the corner.

Pulsation Tectonics

The smooth adherence of cloud to sky,

the boxcars are lined up and edging out.

Barns are the subjects of drawings.

A carving is strong and just a carving in stone.

Berries ripen in the hand.

The simple bed is too soft.

Morphology

Hills of drape, the new roads wind through.

The wagon wheels and the small

of the road is upon us.

You say wagon wheels and I criss-cross.

I sample the doubles; there is a way to write out what we mean.

The hope that comes is terribly old.

The water will not even sell. Bricks in buckets, broken.

Brave men straighten their sport coats in the forest that craves shade.

Rectangular Coordinates

The boxer feels the tide and the elevation.

The wife wheels in.

She lives in this box.

Circle of **Ascension**

The dull mind narrows in the middle of spring.

An ice limb cracks outside the window. Stone-light and sliver.

 Stone-square, the boulder at the corner blocks the view. We can trust

the crowd that moves rigid over the hill, across the drill field.

Fourth month. How the braids of the girl in front of me unravel.

The sides of the cold faucet condensate.

The grass is coarse against our faces.

III

Mother

At breakfast you tell me about
a coyote pack in the city,
the park the last place one was spotted,
just past the bend in the river,
its tail fuller than a dog's,
its ears large and shaped different
than a dog's.
Over the telephone this evening
you tell me details on the wedding
of the prince, there
will be protesters, the tornado
in Iowa took a town,
there's a plan to clone
and mass-produce colossal redwoods,
the tallest living things on Earth.

Loneliness of a Kind

Her heart felt small as the tea light.

The doors opened wide onto the green lawn

but she knew her true friends were too large

to enter the house, a thumb alone taking up

the kitchen island as it is rolled into place.

She may never have to marry again, the papers from

her first marriage in a case too tiny

even for tweezers to break open.

She was carting a calf from the back patio

when the rain began.

Golden Meadow

In the envelope from the old store
we found three small pewter businessmen.
The businessmen were groomed
and wearing suits. We put them
in the window and then it was like we were
watching a movie. The men were standing,
gathered there, we agreed,
to talk about a serious situation.
We turned one so his back was to us
so we could not hear what he was saying.
The one facing us had a painted mustache
and he was the only one smiling.
We decided their wives were at home.
We sat on ice chests in the dark.
In the next scene they were salesmen.

Nighttime in Sequence

Her earring is a live lobster but its claws are fixed.
"Sidestep to the bar," he says to her, and she does,
the trivial nature of their relationship is irritating to her
and their children are not as well adjusted as they should be.
They act up in almost all public situations. Their teeth and mouths
are red from eating popsicles. They are losing the chase game
outside of the large house. The kingdom is only a sand castle.

Gulf of Lions

This margin, you say, *has been created* by an Oligo-Aguitanian rifting and followed by oceanic accretion in the Provence Basin during the Burdigalian. *This margin,* you say, *has been created,* during the Burdigalian, The Messian Event, a clear marker within the history of the basin. *This margin,* you say, *has been created,* the initial period of margin foundation, from beginning continental extension to oceanization. *This margin,* you say, *has been created.* You propose a model for the formation of the Gulf of Lions in three steps: the first, a thermic event, the second, a rupture, the third corresponds to the formation of typical oceanic crust in the center of the basin. *This margin,* you say, *has been created.*

If We Are Very Still

We might hear the wind. We might hear the barges come closer than ever, the river waking against the rocky bank. Our streets will fill with rose water. We can tell time. The dam will close. We might hear the children repeating the day's math assignment, each with a pencil in air, the papers before them curling into soft horns.

Raccoon Point

What's left of the thinned handkerchief, a vine island,

this elongated frown on him. The cap, in these confines

even in lowest of trees,

halved and spread, this dotted heart-shape now on Caillou Bay, we can
    see if we squint.

The most western edge, the lonely calf, the last triangle sail.

The strand of hair in a fist. Someone tells someone the story

of the child's mother and the child's mother tells the story to the child.

Remnant, the tail, the most remaining. The sun's odd fashion of pink
    slapped

on the side of the house just before dark.

We wait until the trailing breath to lose—a kind of death.

We are all here on the last of last island, shaking our coats of fur after a
    swim.

At the Coastline

Becky blasted us with bare nets.

Her shifts were limitless and grassy.

When she first appeared we thought

she was ghostly, a reversal of herself.

Her light hair was stippled with blue.

She understood us only by reading the way

our mouths moved.

We held our arms out

under the cloudy sky. Her skiff leaned.

Holding the Pose

Inside the odd camera, she wondered, could the picture be priceless?

A crack in the lens makes the face

look like two.

Her hand is held straight as a mini saw.

Brand new brackets hold her legs in place—she is able to stand.

Modeling makes anyone tired. The rawness of her lips is due to the wind.

He never knew she would cry like a cat.

He never knew her edginess would outweigh her loose handshake.

The silver bags in her borrowed car keep the money safe.

Her body is already positioned, one leg crossed over the other,

her arms arranged like slender baguettes.

A Worry

The girl in her den thinks

this tornado embedded

is too large to fit in her yard

but she worries that without it

she will never know what the funnel

is capable of, will never watch cows,

a house, a picket fence, a girl

in her den thinking this

tornado is embedded is too large.

The Flashlight Shines Bright Into the Living Room Window

Her long hairstyle is loose and falling from her neck as its pins are old.
My daughter is older than the boy in the closet
and so I keep her out and fully dressed in dress and shoes and tights
and reading glasses.
She is having coffee with her neighbor friend, a girl with short blond
hair wearing a soiled apron.
The quiet in the house as the sea level rises is the kind that remains solid —
until the girls get up for a walk about the room
and the boards under the flimsy floor bend and slant.

Cheniere Caminada

We see it

form a terrible trough

and hear that ringing

we have been told about.

We hold a map and we

know where the northwest

Caribbean Sea is. We think

of the sky as a hood. We walk across

rice fields and wade in calm, cold water.

We walk between orange trees.

We watch the shrimp cannery slipping and tilting

into the water. The trees bend.

We hear the waves in the evening.

We fall beneath the magnifying surge.

We come up, hold onto a fence, move in mud.

We wait for a day in a church with the priest

and a light, read a book on United States presidents.

Island Road

No way to ride. No blank sky on the edge. The scissors are there for making skirts. No bowl of beans. Parts. The lock. The bridge of concrete. The rabbits are so small they disappear in the marsh grass. No pears. No position for or against. The families are fewer. No means by which to measure the depth. No weeks. The fourth finger is bent. Cool pan on the stove. Carved face. Blue waters. Naming animals is easy.

History of Wetland Loss

At dusk we fell into our little straw beds
having digested hanger steaks and pasta elbows.
The buds surprised us with their pink.
If we held our eyes closed for long enough
we would see them bloom again.
The architect told us to imagine rooms with light
and sharp angles. Underwater, we knew
that it was best to hold our breath.
We found ourselves closing our eyes
and our mouths for longer and longer
spells of time, and our heads felt weak.
This anniversary was predicted
just as all others, the gifts we received
wrapped in white paper and tied with string
like blocks of cheese at intertidal elevation.

Tide Gauge

In her heart she keeps a lion. In her hair there is a medallion. Her brother wears a suit. It is here that she learned to spell, to add, to make a pie. The dogwood's flowers are brighter than paper. She can see them. She learns to swim in the pool at the park nearby. She holds a carving up to the sun and for a moment her face is in the shadow. She keeps her eyes closed so that she can think of the shape a while longer.

Unmasking of Fault Displacement

We're waving at the man driving the boat. We've already lied to the yardman. We're satisfied with the knee replacement. We'll try to play a card game to win. We've taken the long road to the Gulf Basin. We've lived without hair. We've been lied to. We've had too many doctors tell us their own problems. We're arguing with the bagger. We'll be fifty next week. We'll have our booster shots. We've won the game we were playing when we had to stay home while the city was shut down. We fled to see a movie. We wiped our hands after bathing the dog. We showered at a campground. We held a fragile art project. We shoved paper into the bin. We know that monkeys are adept at picking up social cues.

To Step Lightly

A cocktail in winter on the ship
shook like a small ocean itself.
The waiters were in black in a row
on deck. The blue water was broken
far out. They came here to marry,
to pretend they had children,
to dance with space between with them,
their arms arched like bridges.
Or a grave in the snow,
without a headstone,
then flowers are placed.
How the fullest moon
looks thin when the words,
in a language you don't understand,
cause someone you cannot see
to laugh. A blast from a transformer
shakes the ground.

Phosphorus

Before a nap, I think about what you might be thinking.

I am thinking of what must be the enormous diet

of the three Burmese Mountain Dogs I saw today.

I am thinking you might be thinking of the nitrogen group,

or maybe the atomic number of phosphorus.

I am thinking you might be thinking of the difference between

white phosphorus and red phosphorus—white phosphorus emitting a faint glow.

I think phosphate minerals are fossils, but I am not sure.

I want to ask you what makes white

phosphorus glow,

if you have ever seen the glow that persists under pressure.

IV

Grand Isle

Her little coat is too small. She's a fool. The wind catches under her skirt and lifts it an inch. Bulls and eggs were in her dreams but she cries this morning knowing it might be a prank. Even the stationary trees might really be moving. The man at the filling station wears a tie that looks like a mouse. The billboard on Rue Abel shows a bull that changes with the next set of panels to a car dealership advertisement. She thinks that her arm itches but her skin plays tricks on her, too. Her brother pretends—or so she thinks—to be her brother for just a moment. He takes a bat and holds it up in the shade.

Mangrove

We follow hatch marks in mud, mark our way with logs and cups,

race crabs to their sink holes.

We hide in the shrubs from boaters, hop the wakes. We come up from
    the green

with our heads in shade, our shoes slipping, our knees in mud.

You tell me level-growing roots twist upwards and downwards,

the upward twists emerge on the water surface.

You tell me if I sink too deep to grab onto the level-growing downward

twists of roots, and pull, pull on the ones that appear on the surface.

Enigma

On my hand two spots have appeared. Was it the scissors?

I talked to a man at the post office about the strange weather. It is so cold.

We agreed we are all upset over the current state of finances,

the rates of exchange, the price of gas.

The girl with the stick in her hand running slowly through the park

in front of my house looked like a thin kite with no string trailing behind.

I had a few errands to run this afternoon, the pharmacy, the bank.

My brother called to tell me his old friend John had shot himself in the head.

My sunglasses fogged up. I said I was so sorry.

I thought of us all at the beach in the 80's.

A heavy snow fell today in the northeast. As I pulled the trash cans out

this evening for tomorrow's pick-up, I stepped over a curled up snakeskin

on the driveway.

Barrier Island, Offshore Bar Theory

In fourth grade we learned about Élie de Beaumont. In 1845 he had the idea that waves moving into shallow water churned up sand, which was deposited in the form of a submarine bar when the waves broke and lost much of their energy. As the bars accreted vertically, they gradually built above sea level. He theorized on barrier islands and also mountain ranges. His largest contribution to science was a geological map of France.

Barrier Island, Submergence Theory

In Golden Meadow a man sits at a bar to drink

thinking about the pieces of his house he will save

from 1853 to 1978 two small semi-protected bays now link

behind barriers open-water areas marshes sink

formation a stable sea level an unchanged pave

in Golden Meadow a man sits at a bar to drink

coastal ridges separated from the mainland by lagoons kink

former lobes reworked by action of wave

from 1853 to 1978 two small semi-protected bays now link

he thinks of what to keep--the chest or plow or piano or bell or mink

the geologists can explain mountain and island and cave

in Golden Meadow a man sits at a bar to drink

just before the storm the sky is always wide and the waters pink

forming beach ridge complexes in the seas' lave

from 1853 to 1978 two small semi-protected bays on the brink he decides to save the calligraphy drawing of ink

but in the leaning houses wait those who will be brave

in Golden Meadow a man sits at a bar to drink

behind the barriers open-water areas marshes sink

Barrier Island, Spit Accretion Theory

In a bucket we have the perch.

He tells me his teacher told him

that barrier sediments come from

longshore sources and that a geologist said this in the 1800s,

an American, he says, and that the sources are sediment

moving in the breaker zone through agitation,

agitation by waves in longshore drift.

Longshore drift is the transportation of sediments,

his sister says. Longshore drift construct spits, he continues.

A spit is a *deposition landform*, his sister says. Longshore drift

construct spits extending from headlands parallel to the coast.

Headlands are *points of land*, his sister says.

He says his teacher said subsequent breaching of spits

by storm waves form barrier islands.

Headlands are sometimes called *heads*, she says.

Or sometimes *a cape*, he says. If it's large, she says.

Isle de Jean Charles, 1976

Cattle graze on acres of green, game trapped, children run.

For one hundred fifty years, on this narrow ridge, like a ship,

with wide dry decks, between Bayou Terrebonne and Bayou Pointe-aux-Chenes, they have lived.

The sun is fat over the water. They keep their fish in floating nets.

On the edge of the island, a piling marked by a white strip of fabric flaps like skin.

ACKNOWLEDGEMENTS

Grateful acknowledgement is made to the following publications in which some of the poems herein first appeared, sometimes with different titles and in slightly different form: *The Cincinnati Review*, *The Southern Poetry Anthology, Volume IV: Louisiana*, *New World Writing*, *The Louisville Review*, *Poets of Living Waters*, *Sentence*, *The Hollins Critic*, *ISLE: Interdisciplinary Studies in Literature and Environment*.

This book was designed and set in Palatino Linotype by RHWD Industries

Cover graphic : The water/land boundaries in southeast Louisiana. Courtesy of the Greater New Orleans Community Data Center.

Photograph of the author by A. M. Barthelme

Printed by Salem Printing

groundhog
POETRY PRESS